YOUR KNOWLEDGE HAS VALUE

- We will publish your bachelor's and
  master's thesis, essays and papers

- Your own eBook and book -
  sold worldwide in all relevant shops

- Earn money with each sale

Upload your text at www.GRIN.com
and publish for free

Frank Machens

Leck Air Base. Conversion of Military Property as a Chance for Renewable Energies

GRIN Verlag

Bibliografische Information der Deutschen Nationalbibliothek:

Die Deutsche Bibliothek verzeichnet diese Publikation in der Deutschen National-
bibliografie; detaillierte bibliografische Daten sind im Internet über http://dnb.d-
nb.de/ abrufbar.

Imprint:

Copyright © 2014 GRIN Verlag GmbH
Druck und Bindung: Books on Demand GmbH, Norderstedt Germany
ISBN: 978-3-656-68087-1

This book at GRIN:

http://www.grin.com/en/e-book/275174/leck-air-base-conversion-of-military-proper-
ty-as-a-chance-for-renewable

Topics of Environmental and Resource Management

Conversion of Military Property:
A Chance for Renewable Energies at The Example of Leck Air Base

University of Southern Denmark (SDU)
Faculty of Business and Social Science

Esbjerg, 25.05.2014

Presented by:
Frank Machens
M.Sc. Environmental and Resource Management
2nd Semester / 4th Quarter

Table of Contents

Abbreviations

DSG	The German Solar Energy Society
DWD	Deutscher Wetterdients – German Weather Information Service
EEG	Erneuerbare Energie Gesetz - renewable energy law
ERM	Environmental and Resource Management
GDP	Gross Domestic Product
GDP PPP	Gross Domestic Product Purchasing Power Parity
EU	European Union
et al.	Et alii (Latin)- translation: and others
n.d.	not defined
PV	photovoltaic
RAF	Royal Air Force
RE	renewable energy
WFG NF	Wirtschaftsförderungsgesellschaft Nordfriesland mbH

Units

m	million
bn	billion
tn	trillion
ha	hectare
km²	square kilometer
p	population
p/km²	population per square kilometer

Abstract

This paper copes with the topic of military conversion at hand of the lead example Leck Air Base. This former military airport of the German Forces is located in Schleswig-Holstein, Germany. The aim of this paper is to outline the chances for the integration of renewable energy based concepts. Hereby, the special, manifolded problematic of conversion processes but also solutions are discussed.

The first introductory chapter illustrates the basic idea of this paper together with an outline of the Leck Air Base conversion project and a definition of relevant terms. The second chapter gives explanations about the region, involved municipalities and the special linkage to the Southern Denmark Region. This is containing information about geographic and socio-economic relations. A deeper economic analysis and an analysis of the overall energy potential is following in the third chapter. Furthermore, case studies of similar projects in the region of Germany are analyzed in the subsequent fourth chapter. Both, the third and the fourth chapter, built up relevant information for the fifth chapter. Here, the results of both chapters are combined, problems are discussed and possible solutions introduced. The final sixth chapter summarizes the results of this paper.

The results leads to specific suggestions for the conversion of Leck Air Base and to more general suggestions for the overall conversion process. This paper shows that the usage of mixed concept out of renewable energy park, industrial park, tourism and natural reserve is indicated for the example of Leck Air Base, but also explains that due to structural, economic characteristics a focus on renewable energy concepts alone is not of advantage. General problems as the resistance of affected residents, especially through the suggested lead technology of wind energy, can partly be avoided by alternative financial concepts as the *Bürgerpark*. Still, it is shown that conversion projects are facing highly individual problems and an assessment on a general basis is not sufficient.

1.Introduction

This paper is written as part of the elective course Topics of Environmental and Resource Management in the second semester of the Master Program Environmental and Resource Management (ERM). The course allows candidates to conduct research in a chosen field under guidance of a professor or lecturer of the Master Program.

In this paper the focus will be on military conversion processes, the resulting chances and challenges for renewable energy systems. Hereby, the practical project of Leck Air Base in Germany will be taken as a lead example. The project itself and certain hold-backs will be introduced later in this introductory chapter, together with relevant definitions. More detailed information about the base and region will be given in the subsequent second chapter. This will include a gradual description of the conversion area itself, the surrounding municipalities, the federal state of Germany the base is located in, and the special connection to the region of Southern Denmark. The third chapter will analyze necessary parameters for the usage of renewable energies, the energy potential for the different sources, and economic factors relevant for the planing process. As this paper aims to point out the chances for renewable energy systems within the conversion of military property, in order to lead to new practical and valuable usage, this chapter is of major importance. The economic analysis will give more detailed information about the regions structure and reveals economic potentials, which might give further impulses for the planing process. The fourth chapter will introduce and discuss case studies from comparable conversion projects. This will give a benchmark for solutions and the opportunity to learn from previous obstacles. In the fifth chapter the results of the analysis of the case studies will be transferred onto our lead example, Leck Air Base. Eventually, the overall results will be summarized in the sixth chapter.

1.1. The Project

Leck Air Base is located in the federal state of Schleswig-Holstein, Germany, between the municipalities of Leck, Tinningstedt and Klixbüll. A closer description of the location can be found in the subsequent second chapter. Communication with several responsible persons showed that the conversion project of Leck Air Base is still in the very beginning of planning. The main idea for the conversion of Leck Air Base is to transform the former military area into an area with following purposes:

- renewable energy park

- industrial park

- innovation center

A clear plan for the main renewable energy source at this phase of planning not yet defined. Likewise, it is too early to define a concept for the industrial park. The lead idea here is a holistic concept, which asks for a renewable energy related industrial park concept.

The main reason for the slow progress are natural protection imposts. The responsible persons are standing in contact with the federal government in Kiel to clarify these problems. The object of these imposts are rare form of grasses, which are spreading in these unused area. Public discussions argue this grasses are going to be removed by natural transformation processes, which is not topic of these paper. Another problem, as indicated in communication with involved people, might be the kind of idealistic view of the affected people of the region, which is said to proclaim renewable energy production as the only solution. This is not considering the abundance of renewable energy in the north, compared to the high demand in the south, which will be discussed in a later part of this paper.

The main aim of the affected people within in the municipality is to create new jobs in order to replace the 800 jobs lost due to the shut down of Leck Air Base and the referring purchasing and economic power.

1.2. Definition of Conversion

The term 'conversion' in its original form goes back to the Latin word of 'conversio', which means 'revolution', 'turning in a whole circle' or simply 'change' (Latin Dictionary: conversio, n.d.). Of course this is not a sufficient definition for the specific form of military conversion used in this paper. As mentioned by Bläser and Kraus (2008) there is no universal definition within the debate for the term of military conversion. After deeper analysis Bläser and Kraus eventuate in the following definition: "transformation from military to civil usage under consideration of the linked effects". (Bläser & Kraus, 2008) Hereby, four different fields of military conversion are defined:

- armament conversion
- conversion of mobile armament objects
- garrison conversion
- regional and environmental conversion

For the topic of this paper relevant are garrison conversion and to some extend regional and environmental conversion.

As defined by Bläser and Kraus, the here relevant conversion processes are often tightly interlinked. It is said, that closure of garrisons and diminishing of personnel contingent is leading to economic and social upheaval. These processes are simultaneously asking for a new usage of the areas and compensation impulses.

2. Region and Object

The following part will first describe the object of this paper, Leck Air Base, prior to the introduction of the surrounding, affected region.

Fundamental information on Leck Air Base will be given. This includes geographical position, characteristics, expansion of the area and a brief overview of the former usages. The surrounding region will be described by geographical position, characteristics and socio-economic attributes. A more detailed economic analysis will be conducted in a subsequent chapter of this paper.

2.1. The Object: Leck Air Base

Leck Air Base, as a former military base of the German Army (Bundeswehr), is located in the federal state of Schleswig-Holstein, North Germany. The particular region, also known as Northern Frisia, is located close to the Danish border. The areal of Leck Air Base shares ground with the three municipalities of Leck, Klixbüll and Tinningstedt. Further information about the size of the area can be found in Table 1. Map 1 gives an overview of the exact location of the object.

The Air Base can look back on a long history of usages: in the Second War by the Third Reich Germany, in the cold war by the British Royal Air Force (RAF), followed by the new formed German Air Force, after the closure of the area as an Air Base in 1993 by Antiaircraft Units of the German Army until the end of 2013. After closure of the airfield for military purpose it was partly used by a private small engine flying club. (German Federal Ministry of Defense, n.d.)

According to the society for economical fostering, Wirtschaftsförderungsgesellschaft Nordfriesland mbH (WFH NF), various constructions are on the area: office buildings, barracks, maintenance workshop air shelters, bunkers for airplanes, and the airfield itself. The buildings are said to be in good condition, and the maintenance workshop is on a high technical niveau, according to the WFG NF (2014).

Leck Air Base had major importance for the economy of the region by providing employment for about 800 people and the related purchasing power in the region.

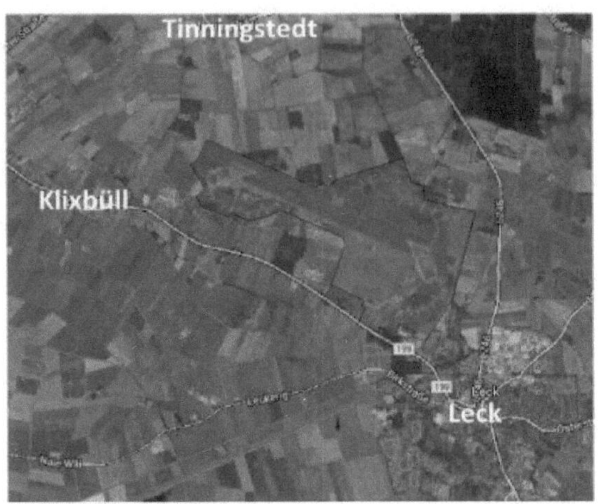

Area of Leck Air Base

Map 1: *Leck Air Base*

(Source: Google Maps, edited)

Area Type	Area [m²] [1]
Base area buildings	55,585
Useable buildings area	54,099
Area traffic ways	349,024
Forest	12,200
Other	2,755,292
Area total	3,226,200 (322.62 ha)

Source:
[1] Society for Economic Fostering Northern Frisia
  (WFG NF)

Table 1: *Leck Air Base - Area*

2.2. The Region and City of Leck

In this part relevant regions as the federal state of Schleswig-Holstein (Germany), the administrative district Northern Frisia, the main municipality of Leck and the region of Southern Denmark will be described. Hereby, relevant information about geographical position and characteristics, and basic socio-economic relationships will be introduced. A detailed economic analysis will follow in the respective, subsequent chapter of this work.

As already stated in the previous part the Air Base Leck is located in Northern Germany, in the federal state of Schleswig-Holstein. Schleswig-Holstein itself is divided into eleven administrative districts, of which Northern Frisia is the relevant one for this paper. (Government of Schleswig-Holstein, n.d.) As mentioned the Air Base shares ground within three different municipalities: Leck, Tinningstedt and Klixbüll. A look at Table 2 reveals the relevance of the municipalities. With about 7600 inhabitants the municipality of Leck as a higher importance referring to Klixbüll with about 950 and Tinningstedt with about 250 inhabitants. Of course it could be argued that, as pointed out, the exact location of the 800 jobs associated to the close down of Leck Air Base is relevant. The impact on the smaller municipalities could indeed be bigger from a quantitative perspective. Still, with reference to Table 2 the overall municipality area of about 56km² is a rather refined area. Therefore, the main municipality of Leck will stand as a representative. Before going into deeper discussion of municipality level, we will turn to superordinated Schleswig-Holstein and Northern Frisia. Schleswig-Holstein has a predestining geographical location between the marginal seas of North Sea and Baltic Sea, with which a coastline of about 411 km is shared. (Government of Schleswig-Holstein, n.d) This becomes majorly important for the later analysis of energy potential. Northern Frisia is the most northerly district of Germany, as can be seen in Map 2. The coastline here has about 156 km, only taken into account the mainland. Which is, compared to 411 km for Schleswig-Holstein, a rather big share. It can be argued to consider the neighboring district of Schleswig-Flensburg. Since Leck Air Base is locate in Northern Frisia and Schleswig-Holstein as an entirety serves as a representative, Schleswig-Flensburg will not be separately observed.

Before turning to further information from Table 2, the relevance of the Southern Denmark Region will be elucidated. Both regions, Southern Denmark and Schleswig-Holstein, are interlinked in various ways. Very obvious is the mere spatial connection, which comes together with a strong historical linkage between both regions. During which the affiliation of nowadays region of Schleswig-Holstein has been changed, which finally lead to annexation by Prussia in the middle of the 19th century. (H. Holborn, 1982) This, among other reasons, explains the variety of languages and minorities on both sides of the nowadays Danish-German border. A multitude of economic initiatives (Government of Schleswig-Holstein, 2010) and the mutual importance of commercial relations indicated the a tight linkage between the regions. With an export of about €1.4bn (7.8% of total) and an import of €2.9bn (14.3% of total) Denmark is the most relevant economic partner for Schleswig-Holstein.(Statistical Division North, 2014) Additionally, a strong linkage through tourism industry is worth mentioning.

Country - Region - Municipality	Area [km²] [1,2,3]	Population [2,3,4,5]	Population Density [/km²] [1,2,3,5]
Germany	357,137.18	80,219,695	229
Federal State of Schleswig Holstein	15,799.56	2,806,531	180
Northern Frisia	2,047.00	162,237	79
Leck	*29.79*	*7,628*	*260*
Klixbüll	*17.44*	*948*	*54*
Tinningstedt	*8.91*	*248*	*28*
Denmark	*42,915.70	5,591,572	130
Syddanmark (Southern Denmark)	12,225.60	1,201,547	99

Note:
 * calculation based on data of previous sources

Source:
 [1] German Federal Statistic Division (2012)
 [2] German Federal Statistic Division - North (2013)
 [3] Statistics Denmark (2012)
 [4] Census Germany (2011)
 [5] World Bank Group (2012)

Table 2:   *Regions - Basic Numbers*

A look at Table 2 gives some basic information about the regions area, population and resulting population density. The referred municipalities have a rather low number of inhabitants as the region of Northern Frisia. The population density of Northern Frisia 79 p/km² is lower as the values for Schleswig-Holstein, 180 p/km², and the overall value for Germany with 229 p/km². Denmark in general has a lower population density, with 130 p/km² compared to Germany; the region of Southern Denmark has with 99 p/km² a lower density as Denmark in general but slightly higher as Northern Frisia. This values indicated that the overall region is rather remote in comparison to it's surrounding.   A statement about the importance of economical sectors can not be made on this values alone. As stated, an economic analysis can be found in a subsequent chapter.

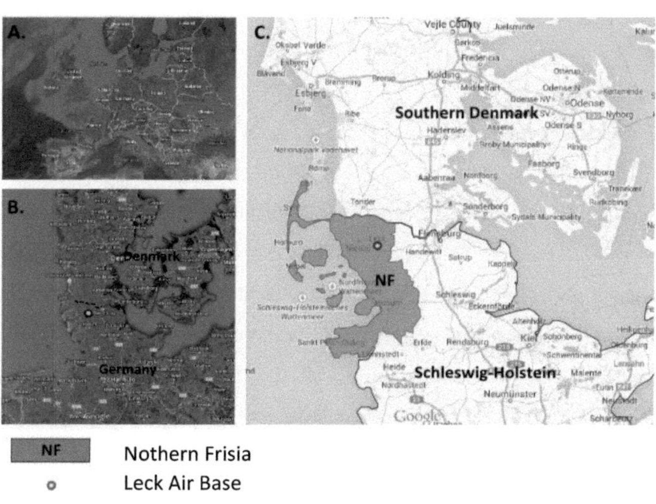

NF   Nothern Frisia

o   Leck Air Base

Map 2:   *Region Southern Denmark - Germany*

(Source: Google Maps, edited)

3. Analysis

In this chapter relevant analysis for the subsequent solutions will be conducted. The analysis will be divided into an economical and energy related part. The economical research will have a look into the economical status and possibilities, as the energy aspects will be analyzed first by potential and second by the current status for relevant regions.

3.1. Economical Aspects

This part will discuss the economical aspects of the relevant regions. Hereby, it is refereed to previous used information as population or population density; additionally, the economic situation is described by gross domestic product (GDP) values and the economic structure. The relevant data to this chapter can be found in Table 3. Germany will be the first object of this analysis before going on subsequent regional levels of Schleswig-Holstein and Northern Frisia. Furthermore, a brief analysis of the economy of Southern Denmark will be conducted. It is refrained to conduct an analysis on municipality level. The reasons are based in the relevance of the municipalities themselves – the small number of inhabitants, the inclusion of the information in the analysis of superordinated regions and the missing statistical data on this organizational level.

As one of the most industrialized countries Germany is the fourth biggest economy with about € 2.7tn nominal GDP according to the World Bank. The GDP per Capita with € 34,126 for about 80m people is one of the highest in the World. (World Bank, 2012) The economic structure shows a typical industrialized distribution according to the classical, economic three sector hypothesis. A rather small primary sector (1.5%) compared to a larger secondary sector (24.7%) and a dominating tertiary sector (73.8%). This values become more important when weighted against the subsequent regions.

Apart from total population and total GDP, Schleswig-Holstein and Northern Frisia share similar economic characteristics. The GDP per Capita for both is about € 28,000, which is about 11.7% below the value for Germany. The economic structure between both regions shows similarities.

Still, in comparison to the values for Germany the primary sector (agriculture, forestry and fishery) is between 46.7% and 100.0% larger. The secondary sector (manufacturing, mining, utility services and construction) is between 14.6% and 20.2% smaller. Finally, the tertiary sector is between 3.9% and 4.7% larger compared to total Germany.

At first Denmark seems to have a substantially higher GDP per Capita as Germany and the associated regions. At this point the usage of nominal GDP and not GDP Purchasing Power Parity (GDP PPP) in Table 3 has to be noticed. This effect can delude the information in this data a direct comparison does not make sense. Still, the comparison of the values within Denmark shows us that similar to Germany and its here relevant regions Southern Denmark has a 14.6% smaller GDP per Capita. The economic structure of Denmark in total shows a larger primary and tertiary but a smaller secondary sector. In Southern Denmark this impression is corrected to some extend. Although the primary sector here is about 52% larger compared to Denmark in total, the secondary sector is 23.9% larger and the tertiary sector is 7.5% smaller.

Summarized it can be said, that the region of Schleswig-Holstein in particular Northern Frisia and Southern Denmark shows some structural similarities. Especially in comparison to the superodinated countries a smaller GDP per Capita might indicate a structural weakness in these regions. The data for economic sectors revealed a stronger focus on the primary sector (agriculture, forestry and fishery) in the relevant regions of Germany and Southern Denmark. Also a stronger secondary sector (manufacturing, mining, utility services and construction) in Southern Denmark has to be noticed. In the relevant regions of Germany the secondary sector is substantially smaller compared to the rest of the country. This vows for a more agricultural based economy in the relevant regions of Germany and supports the hypothesis of structural weakness.

Country - Region - Municipality	Nominal GDP [mil €] [1,2,6]	Nominal GDP per Capita [€/capita]*	Economic Structure (by employment) [2,7,8,9]			Population [2,3,4,5]	Population Density [/km²] [1,2,3,5]
			primary	secondary	tertiary		
Germany	2,737,600	34,126	1.5	24.7	73.8	80,219,695	229
Federal State of Schleswig Holstein	77,275	27,534	3.0	19.7	77.3	2,806,531	180
Northern Frisia	4,681	28,853	2.2	21.1	76.7	162,237	79
Leck	-	-	-	-	-	*7,628*	*260*
Klixbüll	-	-	-	-	-	*948*	*54*
Tinningstedt	-	-	-	-	-	*248*	*28*
Denmark	249,125	44,554	2.5	19.2	78.3	5,591,572	130
Syddanmark (Southern Denmark)	45,717	38,048	3.8	23.8	72.4	1,201,547	99

Source:
[1] Federal Statistical Office Germany [2012], www.destatis.de
[2] Statistical Office for Hamburg and Schleswig-Holstein [2013], www.statistik-nord.de
[3] Statistics Denmark [2012], www.statistikbanken.dk
[4] Census Germany [2011], www.zensus2011.de
[5] World Bank Group [2012], www.worldbank.org
[6] Directorate-General Eurostat [2010], www.epp.eurostat.ec.europa.eu
[7] Federal Statistical Office Germany [2013], www.destatis.de
[8] Statistical Office for Hamburg and Schleswig-Holstein [2012], www.statistik-nord.de
[9] Statistics Denmark [2013], www.statistikbanken.dk

Note:
* For Denmark and Southern Denmark without Greenland
And Faroe Islands

Table 3:   *Economic of Regions*

In order to give a more comprehensive picture apart from the economic perspective the following paragraph will outline an overview of the most important companies and sectors in the region.

The largest companies of Schleswig-Holstein are situated in the energy industry sector (HSH Nordbank, 2012): Raffinerie Heide GmbH, EON Hanse AG, Orlen Deutschland GmbH. Northern Frisia is known for a strong tourism industry, but also a strong renwable energy sector. Companies as Vestas Deutschland GmbH, Repower Deutschland GmbH, Danisco Deutschland GmbH are among the most important contributors (WFG-NF, 2011). In the municipality of Leck Clausen & Bosse GmbH, a company of the printing industry, is the most important employer (WGF-NF, 2011). In Southern Denmark the energy industry and tourism are well prominent. Some of the largest companies here are: Dong Energy A/S, Danfoss A/S, Lego System A/S. (Content Media Partner Nordic AB, 2014)

3.2. Energy Aspects

Since renewable energies play a lead roll in the concept of the initial project, this section is analyzing several aspects of renewable energies in the referred region. Hereby, the energy potential of wind-, solar- and bio-energy of the region around Leck Air Base is considered, and additionally compared to the current status of usage. Germany and the here relevant regions of Schleswig-Holstein and Northern Frisia will be in the focus of the conducted research. Denmark and the region of Southern Denmark are not of major importance here with reference to the actual location of the object, Leck Air Base.

3.2.1. Energy Potential

At first the focus is drawn to the general renewable energy potential of Germany, Schleswig-Holstein and Northern Frisia. Hereby, the energy potential of PV-solar, wind and bio will be discussed in that order. Map 3 Is of major importance for the subsequent section.

The PV-solar energy potential is affected by multiple factors. A general formula for the yield of an PV-solar system is given by A. Goetzberger and V.U. Hoffmann (2005):

$$Y_F = Y_R \times PR \times SH_F$$

Y_F: final yield
Y_R: irradiation at solar module area during considered time [kWh/m2]
PR: Performance Ratio
SH_F: factor for unavoidable shading

Without going to much into technical detail, PR represents technical features of specific PV-solar-modules, which is of no relevance for this paper. SH_F represents shading caused by different sources as pollution, buildings, or others. The usage of the abbreviations and structure of formula may vary in different sources. Pollution for instance can be of importance for modules installed in densely populated areas, as fog and clouds can be in coast near areas. Still, the main defining factor is represented in Map 3A., the irradiation at solar modules. For Northern Frisia it is between 1100kWh/m² and 1200kWh/m², which is compared to the southern areas of Germany with occasional spots of 1300kWh/m² rather low. As can be seen in Figure 2, which will be discussed later in more detail, solar energy makes a small part of the energy mix of Schleswig-Holstein, with about 3.2%. This could be taken as an indication pointing to more efficient methods.

The major factor for wind energy potential can be found on Map 3B., the wind speed. As can be seen in E.Au (2006), the common formula for wind power is given by:

$$P = 1/2\ \rho v^3 A$$

P: wind power
ρ: air density
v: wind speed
A: cross sectional area

Here it the velocity of the wind in the power of three clearly shows the main defining factor of the overall potential. Map 3B. shows wind speed between 4.6m/s and 6.2m/s for the region of Northern Frisia. In comparison to the overall values these values are the highest for Germany as the map indicates. This is represented as well in the energy mix of Schleswig-Holstein, Figure 2, with about 23.9%.

The sources for the production of bio-energy can be divided into four categories (The German Solar Energy Society, 2005):

- Energy crop
- post harvest residues
- organic by-products
- organic waste

The potential of bio-energy in a region is directly related to the availability of these sources. Therefore the first industrial sector, which has been object to a previous analysis, has major importance. Compared to overall Germany the region of Schleswig-Holstein, Northern Frisia and also Southern Denmark - as a potential source - has an increased primary industrial sector. The potential can therefore be seen as good. Unlike as for wind and solar energy, the production of bio-energy often need several locations. The origin of the source for bio-energy seldom is the location of transformation or energy extraction. For the project of Leck Air Base this implies several different possibilities. The area can be used for cultivation of source, for transformation or energy extraction. Due to the previous mentioned restrictions up on the area and the size of the area the usage as a cultivation location is non recommendable. From a holistic point of view only small systems could be applied here. Figure 2 shows the significance of bioenergy for Schleswig-Holstein, with 9.6%.

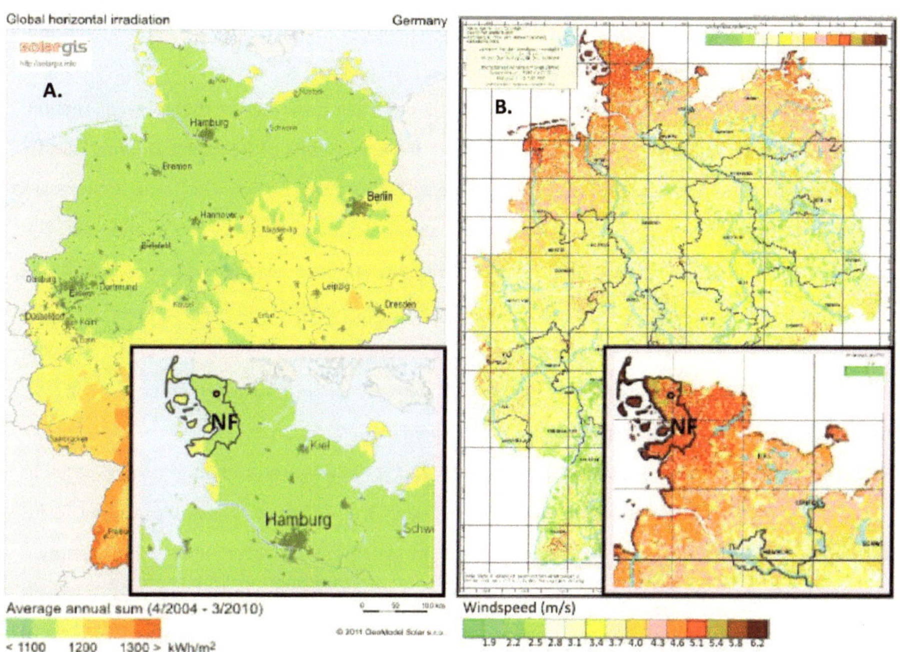

●   Leck Air Base

NF   Northern Frisia

Map 3:   *Solar and Wind Energy Potential*

A. Solar Energy Potential
B. Wind Energy Potential

(Source: A. GeoModel Solar, www.solargis.info, [2011], edited
B. Deutscher Wetterdienst (DWD), [2004], edited)

3.2.2. Current Status

This part will describe the nowadays status of renewable energy in the regions of Schleswig-Holstein , Northern Frisia and the relevant municipalities. Hereby, the energy-mix and the already installed capacities will be analyzed.

To some extend the content of figure 1 and figure 2 has been described in the previous section. To give a more complete overview of the situation, the energy mix of Germany, Denmark and – more importantly – Schleswig-Holstein will be described. As can be seen in figure 1 there are some noteworthy differences between Germany and Denmark in terms of the energy-mix. The main energy source in Germany with about 37.5% is coal and peat, followed by a still strong nuclear energy dependency with 22.7%., despite the planed total shut down in 2022 (Bundesregierung.de, 2011). A strong renewable energy part of 27.1% should be noticed. Denmark has a more homogeneous energy-mix with the main source of crude oil (53.5%), natural gas (30.1%) and the rest produced by renewable energy sources (16.4%).

In reference to renewable energies, Germany with 27.1% has a higher participation compared to Denmark, with 16.4%. Additionally, the ratio between the different renewable energy forms shows that Denmark seems to focus more on geothermal, solar and wind energy.

Figure 2 highlights the differences in the energy mix between Schleswig-Holstein and Germany. Schleswig-Holstein depends less on coal and peat (12.4%), therefore more on nuclear (44.8%) and renewable energies (36.8%). A closer look reveals the importance of wind energy with about 23.9% of the total energy mix, followed by bio-energy with 9.6% and solar with about 3.3% of the total energy mix.

The DGS (n.d.) published an additional source in form of a databank. Here, on the basis of the requirement for publication by network operators energy data from RE is summarized. Although there are some concerns regarding the overall quality of the source, stated by DGS, it can be seen as a complementary tool. Extrapolated data to energy consume is compared to energy production out of renewable sources. The relevant information is summarized in Table 4. Here it can be seen that Northern Frisia covers about 261.1% of its energy demand with renewable energies; it actually surplus in energy production. As the table shows, main responsible therefore is the wind energy with about 699 facilities. The municipalities, which are sharing ground with Leck Air Base, are differing on this account. Here, bio-energy production has the largest share, with about 51.7% of the total RE production. 12 facilities can be found. Wind energy with just three facilities produces about 22.5%. The combined municipalities could cover their energy demand to 94.5% from RE.

	electricity Consumption [MWh/yr]	Solar		Wind		Bio		Total RE production [MWh/yr]	%RE of Consumption
		produced electricity [MWh/yr]	number of Facilities	produced electricity [MWh/yr]	number of Facilities	produced electricity [MWh/yr]	number of Facilities		
Northern Frisia	1,237,864	300,556	6,793	2,424,401	699	517,373	199	3,242,330	261.9
Single Municipalities									
Leck	56,580	6,008	226	1	1	20,948	8	26,957	47.6
Klixbüll	6,993	10,052	42	13,848	2	10,287	4	34,187	488.9
Tinningstedt	1,502	366	6	0	0	0	0	366	24.4
Total Municipalities									
Electricity	65,075	16,426		13,849		31,235		61,510	94.5
Number of Facilities			274		3		12		

Table 4:   *Renewable Energy Production and Number of Facilities*

(Source: DGS, www.energymap.info)

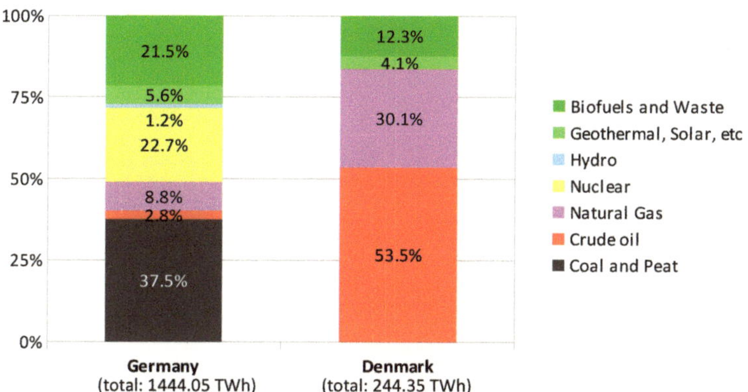

Figure 1:   *Energy Production in Germany and Denmark*

Note: Hydro Energy with a value under 0.1% is not represented for Denmark.
(Based on: International Energy Agency, www.IEA.org, [2011])

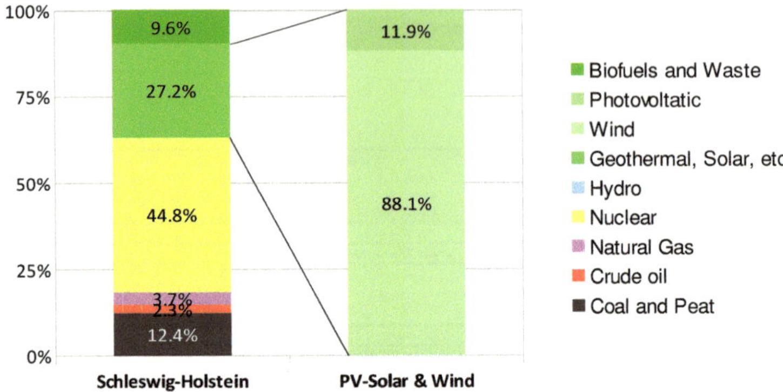

Figure 2:   *Energy Production Schleswig-Holstein*

Note: Hydro Energy with a value of below 0.1% is not represented
(Based on: Ministry of Energy, Agriculture, the Environment and Rural Areas
          Schleswig-Holstein, Report of Federal Government – Print 18/889, [2013])

3.3.Summarized Results

After the analysis of the economical and energy related background, the results will briefly be summarized.

First the economical results will be presented. Hereby, it has to be mentioned, that the results at first are seeming to have some fundamental downsides. But on the other hand are offering some opportunities as will be pointed out. Following economic results regarding the region around Air Base Leck, in reference to the appendant countries, can be summarized:

- lower GDP values
- differing structure of economic sectors: larger agricultural sector, smaller industrial sector
- lower population density
- largest companies out of energy producing industry
- strong economic and historical linkage between Schleswig-Holstein, Northern Frisa and Southern Denmark

As mentioned above some of these economical aspects bear advantages for the potential of renewable energy. The energy potential characteristics have been restricted mainly to the directly affected regions, and less to Germany and Denmark. The results can be summarized as following:

- substantial wind energy potential
- some solar energy potential
- substantial bio energy potential

For all energy forms a low population density and spreading is of advantage. This is due to the availability of more space, and the assumption of reduction of people related conflicts between stakeholders. Bio-energy, as argued, is strongly dependent from the agricultural sector, which has a larger share in the structure of economic sectors as to the compared countries. It has to be mentioned, that the smaller industry and a presumed 'structural weakness' might cause problems. Eventually, the northern regions of Germany seem to have an abundance of RE potential, which eventually asks for a consumer. These consumers, as the heavy industry, seem especially located in the south and west part of Germany (DENA, 2010). Since Germany has a rather complicated energy price market, partly due to the introduction of the German Renewable Energy Act (EEG), this abundance is not reflected in the energy price (Monopolkommission, 2013).

Additionally, the already installed facilities have to be noticed. Bio and wind energy is already widely used as can be seen in Table 4 of this paper.

4. Case Studies

Given the theoretical approach of this work the manifolded problems, which have been described in the introductory part, can not be solved at once. The aim of this work is to give an idea of how a renewable energy based solution for Leck Air Base can be designed. Therefore, this chapter will include previous results of this work and analyze existing projects. These will later be combined with the results of the previous conducted analysis the conversion of Leck Air Base.

The transformation of the German Army (German Ministry of Defense, 2010), lead to the closure of military bases. There are many examples of conversion projects of former military property in Germany. In the following part of this chapter a choice of these projects will be introduced and discussed. Various solutions have been found in former projects as the usage for tourism, integration in settlement planing, natural reserves, transportation industry, culture and education, and as energy parks. The choice orientates itself on the main purpose of the former project, the usage as renewable energy park, innovation and industrial center. Furthermore, the population and regional structure in form of population density have been taken into account.

4.1. Case: Eggebek

Eggebek Air Base was formally used by the German Navy. It is located similar to Leck Air Base in the Federal State of Schleswig-Holstein. After closure in 2006 first plans were made. These included the usage similar as the initial plans for Leck Air Base. The area of about
420ha was planed to host an energy and technology park in combination with a 60ha industrial park. The lead technology supposed to be wind energy in form of a test facility under supervision of the University of Applied Sciences Flensburg, this should be complemented by further but smaller bio and solar energy facilities (Ministry of Economic Affairs, Employment, Transport and Technology, 2010).

The initial plans faced a strong opposition within the population of the municipality of Eggebek.

Main concern seemed to be the effects of the wind energy facilities. The problem could not be resolved in total, which lead to a change in planing. Finally, the priorities of the project had to be adapted. As can be seen in Table 5 the lead technology now is solar energy. With an 83.5MW PV-solar energy facility the former Eggebek Air Base is now one of the largest PV-solar energy parks in Germany and Europe. This is complemented by a 0.75 MW bio energy facility and a 2.05MW wind energy facility for research purposes used by the University of Applied Sciences Flensburg.

The idea of an industrial park was implemented. The initial multiple investors plan had to be changed to a single investor, GPC Gewerbepark Carstensen GmbH. (GPC Gewerbepark Carstensen GmbH, 2014)

The main problem of this promising project was the resistance within the population, which some strong analogy to the *Not In My Backyard Effect* (C.R.Warren et al., 2005). Therefore, the less favorable energy source of PV-solar had to be chosen as lead technology instead of wind energy, compare Map 3. Additionally, a big asset of the property could not be utilized, the airplane runway. Other examples as the airports of Zweibrücken or Frankfurt Hahn indicates that a continuous utilization of this asset is feasible. Although different factors have to be considered during planing as competing airports or population density in the draw area.

4.2.Case: Saerbeck

In 2011 the municipality of Saerbeck took over the area of the former ammunition depot of the German Army. Saerbeck itself is located in the federal state of North Rhine-Westphalia and is a rather small municipality with about 7000 inhabitants. An operating company, SaerPVBioenergieGmbH und Co KG has been founded for administrative purposes of the energy park. The population of Saerbeck has the possibility to participate in the financing of the project; a concept called *Bügerpark*. The total area makes about 90ha from which 25ha are used as a natural reserve. Contrary to Eggebek and the initial ideas for Leck Air Base, here no industrial park was planed. Gradually, PV-solar, wind-turbines and bio-energy facilities have been installed. Unlike from Eggebek the lead energy source here is wind energy. A 5.7MW PV-solar facility has been installed, followed by 2MW bio energy and the main source a 21.3MW wind energy facility.

Accompanied is the energy park Saerbeck by the University of Applied Science of the close by city of Münster. (Municipality of Saerbeck, n.d.) All relevant information can also be seen in Table 5. Following the official information the conversion process here proceeded without major complications, although it mainly relies on wind turbines.

4.3. Case: Morbach

The Energy Site Morbach is based on the former territory of the American Air Force, which gave up the former ammunition depot already in 1995. Morbach is located in the federal state of Rhineland-Palatinate and has a population of about 10,000 people. First plans to transform the area into a tourism facility had to be upset. Finally, in 2001 the idea of an energy park came up. A consortium of investors, which also include the people of the municipality, has been founded. Energy Side Morbach is based on a concept integrating renewable energies, an industrial park and to some extend tourism in form of guided tours and an information center. Additionally, the University of Applied Science Trier participates in small scale cooperation. The industrial park is made of a 23ha area of the total 145ha. The concept so far appears to be very similar to the plans for Leck Air Base. The lead technology for the energy park is wind energy, with a 28MW facility. This is combined with the planed 2.1MW PV-solar system, by now Is having a 0.5MW capacity, and a 1.2MW bio energy facility. (Municipality Morbach, n.d,) Please note Table 5.

Similar to the previous introduced site in Morbach there is no apparent problematic. It seems to be a holistic, functioning design. The region itself is very much comparable with previous regions.

	Eggebek [1][2]	Saerbeck [3]	Morbach [4]	Leck
Population	2200	7000	10000	8900
Area Total [ha]	420	90	145	322
Wind [MW]	2.05	21.3	28	lead technology
Solar [MW]	83.5	5.7	2.1	n.d.
Bio [MW]	0.7	2	1.2	n.d.
Investors	one	multiple	multiple	multiple
Industrial Park	x	x	x	x
University Participation	x	x	x	x
Tourism Approach	-	x	x	x
Natural Reserve	-	x	-	x

Note:

n.d.   not defined

Source:

[1] Ministry of Economic Affairs, Employment, Transport and Technology [2010]

[2] GPC Gewerbepark Carstensen GmbH [2014]

[3] Municipality of Saerbeck [n.d.]

[4] Municipality Morbach [n.d]

Table 5:   *Overview Case Studies*

5.Comparison

This part will compare the results of the economical and energy potential analysis with the case studies of the previous chapter. This process should help to find and outline certain problems within conversion processes to renewable energy parks. Additionally, solutions found in the case studies will be discussed in respect of applicability for Leck Air Base.

The case studies revealed, that there are various of problems with conversion to renewable energy parks, but also solutions. The project of Leck Air Base currently, as mentioned in the introduction, faces problems from an unsolved situation of environmental protection imposts. The case studies, in reference to the example of Eggebek, showed that there are also can be problems faced in a later step of the planing process. Here, it was the problem of the wind turbines, which very much resembles the Not In My Backyard Effect (R.Warren et al., 2005). The sources about Morbach and Saerbeck have not shown any signs of problems, which is not a sufficient prove for the non-existence of those.

The applied lead technology with exemption of Eggebek is wind energy. This underlines the findings of the energy potential and the overall development within the industry. Eggebek focuses on PV-solar energy, because of the described resistance within the population against wind energy. This compromise finally lead to the construction of one of the biggest PV-solar installations in Germany. Certainly, this is not the most efficient solution but still an advantage for the usage of renewable energy sources.

As can be seen in Table 5, all projects include the setup of an industrial park, which makes sense in many ways. Described further above, there is a vast amount of renewable energy available in the north, but just little in the south. This and the current storage problem of the unsteady renewable energy forms, as wind and PV-solar, makes the direct use of this energy in an industrial park favorable.

Likewise, all projects cooperate with local universities. This cooperation has differing intensity in each project. Still, both universities, which are mainly universities of applied science, and the projects themselves seem to benefit from this cooperation.

All projects without Eggebek integrate a kind of tourism within their concepts. Eggebek seems to have a higher priority on the industrial park. Especially in combination with natural reserves this seem to make sense. Hereby, two different targets can mutually foster each other. The natural reserves are often used as ecological compensation areas. Thereby, another aim is fulfilled. For Leck Air Base this also could be valid solution, especially in reference to the nearby touristic areas of the Coast of the North Sea.

The last to mention point from the case studies is the investors structure. Most projects seem to favor financing by multiple investors. Again Eggebek makes a difference. The reason here is to find as before in the change of the initial plans. Especially a finance model called *Bürgerpark* seems to remarkable. Hereby, the affected population or private people in general get the change to invest directly in the energy park. In that way the participation and success of the project is tied to the affected private persons. Of course, this can not be general recipe for the avoidance of conflicts, but might serve as a tool. A multiple investor plan gives more security in the financial planing, also it makes it more complicated.

What are the results for the Leck Air Base project? As pointed out above, the location of Leck Air Base is offering some advantages. It disposes of proximity to one of the most attractive tourism areas in Germany. Of course this is only a small part of the holistic set up. Furthermore, an airfield is available. As discussed, there is various competition in this sector within Germany (Hamburg, Lübeck) but also in Denmark (Billund). Considering these factors, the main import sector still should be the usage of the renewable energy potential in combination with an industrial park. A glance on the energy potential analysis, and in consideration of the built up experience on wind energy in the region the lead technology wind energy is recommendable. A focus should be laid on the development of a solid plan for the industrial park. Here, is were new jobs to a larger extend can be created, which strengthens the economic situation of the surrounding municipalities. For a holistic concept it would of course be recommendable to attract companies from the renewable energy sector. A bigger economical effect might be the attraction of companies from another background. This will avoid *cannibalism effects* within close by regions (Adriano Profeta, 2008). To rely on renewable energy production alone appears to be non-sufficient, as the north-south-incline illustrated. A well balanced mix of renewable energy park, industrial park, tourism and natural reserve under participation of regional universities is indicated.

Still, it has to be mentioned, that the Leck Air Base project has not made it to this part of the project planing. Since, natural protection imposts have to be clarifies. Before a next step can be reached.

<table>
<tr><td colspan="2" rowspan="2"></td><td colspan="2" align="center">Internal</td></tr>
<tr><td align="center">Strength</td><td align="center">Weakness</td></tr>
<tr><td rowspan="2" align="center">External</td><td align="center">Opportunity</td><td>renewable energy know-how in region: Use the knowledge, attract local partners

comparatively large agricultural sector: advantage for bio energy

Strong, regional linkage to Denmark: attract companies and business partners from Denmark

Substantial Renewable Energy Potential: Use with priority 1.wind, 2.bio, 3.solar</td><td>construction imposts regarding old bunkers: for instance a stand for solar panels (compare case studies)

Air Field in combination with weak market stand: small scale airplane clubs, stand for Solar panels (Eggebek)</td></tr>
<tr><td align="center">Threat</td><td>low population density: positive effect on Resistance (NIMBY-effect)</td><td>structural weakness, small industrial sector: create a unique market position

Natural protection imposts: A chance to develop a tourism concept

North-South-Energy-Problematic: attract Energy demanding industries</td></tr>
</table>

Figure 3:   *SWOT-Analysis*

(Source: own design, standard SWOT)

6.Conclusion

This chapter will present the results of this paper in a summarized manner. Furthermore, recommendations for additional studies are made.

The results can be seen as followed:

- mixed concept out of renewable energy park, industrial park, tourism and natural reserve is indicated in the region

- lead technology for the renewable energy park should best be wind energy

- a heterogeneous concept for the industrial park it is recommended. The renewable energies could be considered with a defined quota

- financial concepts as for example the *Bürgerpark* can help to reduce conflicts, but are certainly can not solve conflicts alone

- weaknesses of a side can be turned into strength. Here the impost for natural protection can be utilized for a small scale tourism concept

- conflicts within a conversion project are having a high level of uniqueness, which is depending on regional characteristics and peoples mentality

This paper can be seen as a fundamental study on the problematic in conversion processes. It offers some solutions, but as with the most studies improvements can be made. This study showed that there are various directions in which further studies could contribute. Conflicts as natural protection imposts, faced in the project of Leck Air Base, can hardly be avoided and seem to be a necessary process. Still, here further studies on the specific problematic caused by the object of protection, a rare grass, seem to be useful. This already being quite project specific, further research on suitable companies in could be made under consideration of the *Cannibalism Effect*. This study mostly operates with macro data about energy potential and economic attributes, more detailed measurements and field studies could contribute to more precise results.

References

Aktuelles Sondergutachten. (2013). Monopolkommission. Retrieved May 20, 2014, from
http://www.monopolkommission.de/aktuell_sg65.html

Am Puls der Gezeiten – Wirtschaft in Nordfriesland. (n.d.). . Retrieved May 13, 2014, from
http://www.wfgnf.de/seiten/de/standort/konversionsflaechen/detail/konversions
flaeche.php?we_objectID=56

Bericht der Landesregierung – Zusammenwachsen. (2010). : Government of Schleswig-Holstein.

Bläser, T., & Kraus, F. (2008). Konversionsflächenmanagement zur nachhaltigen
Wiedernutzung freigegebener militärischer Liegen- schaften (REFINA-KoM) . Studien zur
Raumplanung und Projektentwicklung , 4/08, 11-13.

Bioenergiepark. (n.d.). . Retrieved May 11, 2014, from http://www.klimakommune-saerbeck.de

conversio. (n.d.). meaning. Retrieved May 22, 2014, from http://www.latin-dictionary.org/
conversio

dena-Netzstudie II. (2010). Retrieved May 8, 2014, from http://www.dena.de/fileadmin
/user_upload/Presse/studien_umfragen/Netzstudie_II/Endbericht_dena-
Netzstudie_II.pdf

DGS. (2005). Bioenergy systems: a guide for installers, architects and engineers. London: James
& James.

Die Stationierung der Bundeswehr in Deutschland. (2011). . Retrieved May 20, 2014, from
http://www.bundeswehr.de/bwde/Stationierungsbroschuere2011.pdf

Energielandschaft Morbach Die Zukunft gestalten!. (n.d.). Energielandschaft Morbach:
Startseite. Retrieved May 22, 2014, from http://www.energielandschaft.de/startseite/

Federal German Ministry of Devens. (2013). BERICHT ZUM STAND DER Neuausrichtung der
Bundeswehr. Retrieved May 22, 2014, from http://www.bmvg.de/resource/resource/

MzEzNTM4MmUzMzMyMmUzMTM1MzMyZTM2MzEzMDMwMzAzMDMwMz
AzMDY4Njc2NzM3NzMzMDZjNmMyMDIwMjAyMDIw/Ressortberichtl.pdf

Federal Govenmnet of Germany (2011). Der Weg zur Energie der Zukunft - sicher, bezahlbar und
umweltfreundlich. Retrieved May 8, 2014, from http://web.archive.org/web/
20111116042621/http://www.bundesregierung.de/Content/DE/__Anlagen/
2011/06/2011-06-06-energiekonzept-eckpunkte,property=publicationFile.pdf

Gewerbepark Carstensen GmbH (n.d.). . Retrieved May 22, 2014, from http://www.gpc.sh/

Goetzberger, A., & Hoffmann, V. U. (2005). Photovoltaic solar energy generation. Berlin: Springer.

Kuijk, C. v. (2013). Die 100 gröSSten Unternehmen in Schleswig-Holstein 2012. . Retrieved May
13, 2014, from http://www.hsh-nordbank.de/media/pdf/presse/publikationen/
studien/Die_100_groessten_SH_2012.pdf

Hau, E., & Renouard, H. v. (2006). Wind turbines fundamentals, technologies, application,
economics (2nd [English] ed.). Berlin: Springer.

Holborn, H. (1982). A history of modern Germany. Princeton: Princeton University Press. 176.

Ministry of Economic Affairs, Employment, Transport and Technology SCHLESWIG-HOLSTEIN
(2010). Informationen für Entscheidungsträger - Konversion richtig anpacken. Retrieved
May   6, 2014, from http://www.schleswig-holstein.de/MWAVT/DE/Services
Broschueren/Wirtschaft/40konversionAnpacken__blob=publicationFile.pdf

Profeta, A. (2008,). Measurement of Cannibalism Effects in buying experiments using Mixed
Logit   Models - The Exampl e of a new Bran d of the "Fruits of Lake Constance" Associ
ation.   Retrieved April 22, 2014, from
http://ap.wzw.tum.de/fileadmin/pdfs/DP_02_2008.pdf

Standortinformation Gemeinde Leck – „Das grüne Herz zwischen den Meeren". (2011). .
Retrieved   May 13, 2014, from http://www.wfg-nf.de/downloads/
Standortdatenblaetter/Standortblatt_Leck_web.pdf

Stando rtinformation Kreis Nordfriesland. (2011). . Retrieved May 13, 2014, from
    http://www.wfg-nf.de/downloads/Standortdatenblaetter/
    StandortblattNordfriesland_web.pdf

Statistical Division North. (2014). 10. Kapitel: Handel, Gastgewerbe, Fremdenverkehr und
    Dienstleistungen. Statistisches Jahrbuch Schleswig-Holstein 2013/2014 (). Hamburg:
    Statistik Amt Nord.

The largest companies by turnover in Region of Southern Denmark. (n.d.). The largest
    companies by turnover in Region of Southern Denmark. Retrieved May 20, 2014, from
    http://www.largestcompanies.com/toplists/denmark/largest-companies-by-
    turnover/region-of-southern-denmark

Warren, C., Lumsden, C., O'Dowd, S., & Birnie, R. 2. Green on green: public perceptions of wind
    power in Scotland and Ireland. Journal of Environmental Planning and Management, 48,
    853-875.

Welcome to Schleswig-Holstein. (n.d.). Schleswig-Holstein – Schleswig-Holstein. Retrieved May
    22, 2014, from http://www.schleswig-holstein.de/Portal/EN/Portal_node.html

World Bank Group. (n.d.). Worldwide GDP. . Retrieved April 21, 2014, from
    http://data.worldbank.org/indicator/NY.GDP.MKTP.CD?order=wbapi_data_value_2012
    +wbapi_data_value+wbapi_data_value-last&sort=asc